ҲИКОЯИ РАҚАМҲО

THE NUMBER STORY

SMALL BOOK ONE
ENGLISH - TAJIK

*Numbers Teach Children
Their Number Names*

written and illustrated by

MISS ANNA

Early Reader Edition of *The Number Story 1*
Bronze Medal Winner, 2016 Wishing Shelf Book Award

Library of Congress Control Number: 2018902040

Names: Miss Anna, author.
Title: Number story : numbers teach children their number names / Miss Anna.
Description: Portland, OR: Lumpy Publishing, 2018.
Identifiers: ISBN 978-1-949320-04-6| LCCN 2018902040
Summary: The pictures and rhymes present stories which introduce numbers 0-10.
Subjects: LCSH Numeration—English—Tajik--Pictorial works--Juvenile literature. | BISAC JUVENILE NONFICTION /
Languages: English—Tajik
Classification: LCC QA141.3 .M57 2018 | DDC 513—dc23

Publisher: Lumpy Publishing
Website: www.missannabooks.com
Email: missanna@missannabooks.com

Paperback: ISBN 978-1-949320-04-6
Printed in the U.S.A.   1 3 5 7 9 10 8 6 4 2

Мехоҳед номҳои рақамиро ёд гиред?

It is very easy and a lot of fun!

Ин хеле осон аст ва бисёр шавқовар!

Say-along our little jingle

Суруд хонед бо мо суруди хурди мо!

starting from Number One!

Мо оғоз меёбад аз Рақами Як!

1

ONE looks like my one finger.

ЯК

мисли як ангушти ман.

ONE!
ЯК!

2

TWO trails a tail.

ДУ

аз дум кашидан.

A TAIL! ДУМИ!

3

THREE   has bumps.

CE

варам дорад.

BUMPY!   ПАСТУ БАЛАНД!

4

FOUR carries a sail.

ЧОР

амалй як киштй.

A SAIL!
ЯК КИШТЙ!

FIVE is a racing track.

ПАНЖ

ин роҳи мусобиқа аст.

VROOM
БРУМ-БРУМ!
1

6

SIX   curves like a snail.

ШАШ

щатъ кардан мисли

тӯщумшуллущ.

A Snail! ТӮЩУМШШУЛЛУЩ!

7

SEVEN has a sharp angle.

ҲАФТ

як кунҷи тез аст.

BE CAREFUL! IT'S SHARP!

Эҳтиёт шав! Ин Тез аст!

8

EIGHT is rollercoaster rails.

ҲАШТ

роҳи оҳани лағшишбозӣ роҳҳо.

YPEEEE!
YIPPEE!

NINE is a bubble on a stick.

НӮХ

як ҳубобча дар чӯб аст.

A BUBBLE!
ҲУБОБЧА!

TEN   is an eye of a whale.

як чашм аз наҳанг.

WINK!
ВИНК!
HELLO! САЛОМ!

And

Ba

0

ZERO is an empty pail.

СИФР

ба мисли холй сатил.

IT'S EMPTY!
Он Холй!

Thank you for playing with us today.

We had a lot of fun too!

Ташаккур ба шумо барои бозӣ кардан бо мо имрӯз.

Мо хеле фароғат дорем!

We are your Number friends,
Zero to Ten,
Who will be here for you~

Мо рақамҳо шуморо дӯст медорем

Сифр ба Даҳ.

Мо ҳамеша барои шумо ҳастем.

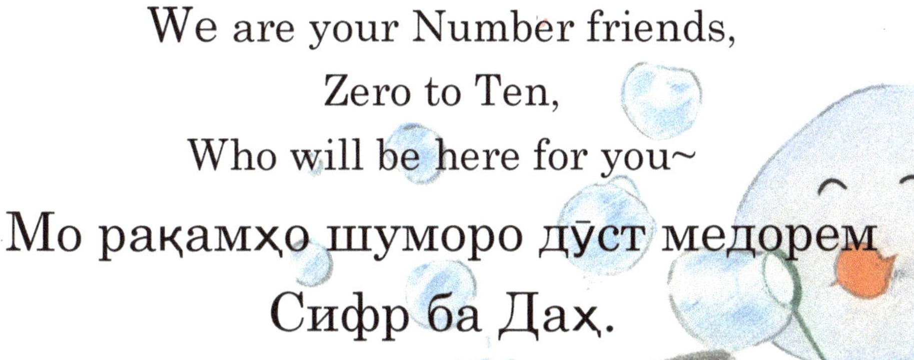

Bye-bye now!
See you again soon!

Хайр!

То дидана!

The Numbers are *SINGING* too!

To sing-a-long, look for Miss Anna Number Story
at your favorite music store like iTUNES.

MP3

Numbers 0-10
IDENTIFYING & COUNTING

Numbers 11-20 & Ordinals
first, second, third...

Numbers 0-100 & Place Values
ones, tens, hundreds...

About Clocks & Telling Time
hours, minutes, seconds

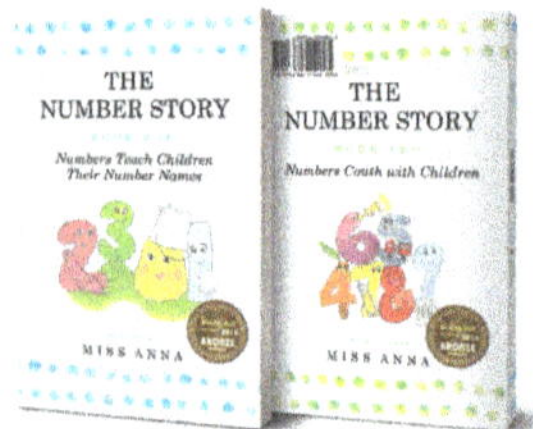

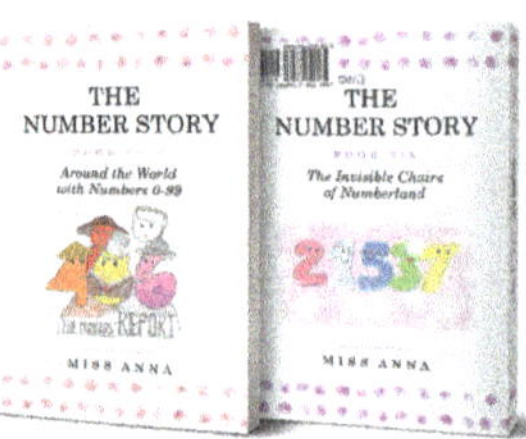

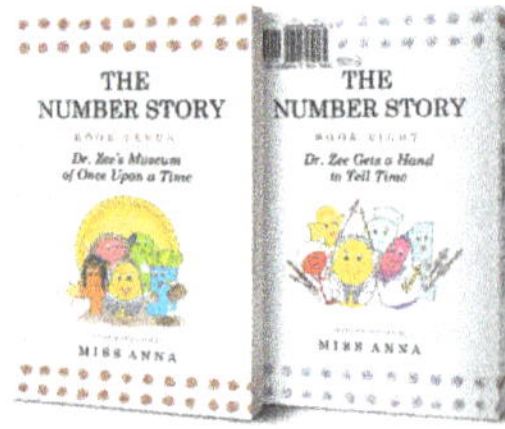

Number Story 1 & 2
isbn: 978-0-996216-48-7

Number Story 3 & 4
isbn: 978-1-945977-01-5

Number Story 5 & 6
isbn: 978-1-945977-06-0

Number Story 7 & 8
isbn: 978-1-949320-40-4

For more Miss Anna books to love,
visit us at

www.missannabooks.com

Numbers are working hard all over the world!
Come Travel the World with Us!